THE POETRY OF TIN

The Poetry of Tin

Walter the Educator

Silent King Books a WhichHead Imprint

Copyright © 2023 by Walter the Educator

All rights reserved. No part of this book may be reproduced in any manner whatsoever without written permission except in the case of brief quotations embodied in critical articles and reviews.

First Printing, 2023

Disclaimer
This book is a literary work; poems are not about specific persons, locations, situations, and/or circumstances unless mentioned in a historical context. This book is for entertainment and informational purposes only. The author and publisher offer this information without warranties expressed or implied. No matter the grounds, neither the author nor the publisher will be accountable for any losses, injuries, or other damages caused by the reader's use of this book. The use of this book acknowledges an understanding and acceptance of this disclaimer.

"Earning a degree in chemistry changed my life!"
- Walter the Educator

dedicated to all the chemistry lovers, like myself, across the world

CONTENTS

Dedication v

Why I Created This Book? 1

One - The Unsung Hero 2

Two - Mankind 4

Three - Guiding Light 6

Four - The Greatest Of All 8

Five - Essence Inspires 10

Six - Longing Restored 12

Seven - Harmony, We'll Be 14

Eight - Cherish Tin 16

Nine - Pure Ecstasy 18

Ten - Debut 20

Eleven - Joys It Brings 22

Twelve - Forever Adored 24

Thirteen - Emotions We Sow	. . . .	26
Fourteen - Tin, Oh Tin	. . . .	28
Fifteen - Uplift	. . . .	30
Sixteen - Revered	. . . .	32
Seventeen - Tin's Magic	. . . .	34
Eighteen - Precious Element	. . . .	36
Nineteen - Never Cease	. . . .	38
Twenty - Single Kiss	. . . .	40
Twenty-One - No Words Can Tell	. . . .	42
Twenty-Two - Sets Us Free	. . . .	44
Twenty-Three - Symbol Of Loyalty	. . . . . .	46
Twenty-Four - Through And Through	. . . . .	48
Twenty-Five - Binding Souls	. . . .	50
Twenty-Six - Eternal Chase	. . . .	52
Twenty-Seven - Love And Endurance	. . . . .	54
Twenty-Eight - The Power Of Tin	. . . .	56
Twenty-Nine - United In Spirit	. . . .	58
Thirty - Come What May	. . . .	60
Thirty-One - Rightful Place	. . . .	62
Thirty-Two - Guiding Embrace	. . . .	64

Thirty-Three - Symbol Of Love 66

Thirty-Four - Enduring And Blessed 68

About The Author 70

WHY I CREATED THIS BOOK?

Poetry has the power to transform even the most mundane subjects into something beautiful and thought-provoking. By focusing on Tin, I can explore its properties, history, and symbolism, and weave them into captivating verses. A poetry book about Tin provides an opportunity to combine science, history, symbolism, and personal expression, resulting in a unique and engaging collection of verses.

ONE

THE UNSUNG HERO

In the realm of elements, Tin does reside,
A metal gray, lustrous, with a humble pride.
Symbolized by Sn, it stands so true,
Tin, the element, with wonders to pursue.

From ancient times, its secrets unfurled,
A treasure sought, a gift to the world.
Malleable and ductile, it bends and shapes,
Crafted into vessels, enhancing landscapes.

Tin, the guardian of culinary delight,
Preserving flavors, sealing them tight.
From cans to foils, it shields our food,
A silent protector, never to elude.

In alloys, Tin finds its perfect blend,
Bronze and pewter, its companions lend.

Strong and resilient, they forge ahead,
Creating beauty, where talents are spread.
  Yet, Tin's story extends beyond its might,
A legend whispered, a tale of light.
The Tin Man, a character, so full of heart,
Seeking emotions, a fresh new start.
  Through battles fought, and friendships gained,
Tin teaches us, love can be regained.
A reminder to cherish, to feel, to live,
For even Tin, has more than it gives.
  So let us honor this element pure,
For its strength, its grace, its allure.
Tin, the unsung hero, a treasure so grand,
Forever it will shine, in the alchemist's hand.

TWO

MANKIND

In the realm of elements, let us sing,
Of Tin, a metal with a versatile ring.
Malleable and silvery, its hue shines bright,
A marvel of nature, a captivating sight.

Tin, oh Tin, in culinary delight,
Preserving our food, day and night.
From cans to containers, you lend a hand,
Keeping our sustenance safe and grand.

Yet Tin, dear Tin, your worth extends far,
For in alloys, you play a vital memoir.
With copper and bronze, you forge a bond,
Creating strength and durability beyond.

In tales of old, a Tin Man we find,
Devoid of a heart, a longing in kind.

Yearning for love, emotions so true,
A quest for a heart, just like me and you.
    For Tin, my friend, you teach us this,
That love and emotions are what we miss.
In a world so cold, we yearn for connection,
To feel alive, to experience affection.
    So let us celebrate Tin, the unsung hero,
A symbol of love, a reminder to show.
That in a world of elements, we find,
The essence of life, the heart of mankind.

THREE

GUIDING LIGHT

In the realm of elements, Tin stands tall,
A metal of value, loved by all.
Its atomic number, fifty, it bears,
With a lustrous shine, it proudly wears.

    Tin, a versatile element of worth,
In preservation, it finds its birth.
From canning fruits to sealing the lid,
Preserving the feast, where flavors are hid.

    Tin, the binder of alloys, so strong,
A marriage of metals, it does belong.
With copper, it forms bronze, sturdy and grand,
A union of strength, forever to withstand.

    Now, let us delve into a different tale,
Of Tin, the symbol of love's longing wail.

The Tin Man, a figure from old folklore,
A heart of tin, yearning for love evermore.
  Oh, Tin, you are more than just a metal,
You symbolize longing and hearts that settle.
You remind us of our essence and core,
The connection we crave, forevermore.
  Unsung hero, Tin, you hold a key,
To the heart of mankind, to what sets us free.
In preservation and love, you play your part,
A reminder of life's essence, from the start.
  So let us celebrate Tin, in all its grace,
A symbol of strength, love, and embrace.
In cans and alloys, you shine so bright,
Tin, the unsung hero, our guiding light.

FOUR

THE GREATEST OF ALL

In the realm of elements, Tin holds its reign,
A metal of luster, with a subtle golden stain.
Its atomic number fifty, its symbol Sn,
Tin, the versatile element, a marvel among men.
    In ancient times, its worth was known,
A preserver of food, a treasure to own.
Tin, the guardian of flavors, the protector of taste,
Preserving sustenance with an enduring embrace.
    Alloys it forms, with strength and grace,
Bronze and pewter, in every place.
A foundation for creativity, a medium to mold,
Tin, the alchemist's muse, a story yet untold.
    But beyond its properties, Tin holds a tale,
Of longing and yearning, of love's epic trail.

In the land of Oz, the Tin Man we see,
A symbol of heartache, of what could never be.
    For Tin, oh Tin, a reminder so clear,
Of the longing for love, the connection we hold dear.
A heart once lost, now replaced with tin,
Seeking compassion, a love that lies within.
    So let us celebrate Tin, in all its might,
A reminder of life's essence, shining so bright.
With preservation and strength, it stands tall,
And the yearning for love, the greatest of all.

FIVE

ESSENCE INSPIRES

In the realm of elements, Tin does reside,
A metal with secrets it does hide.
A guardian of food, a preserver of taste,
Tin's role in our lives we can't erase.

In cans and containers, it finds a place,
Preserving our sustenance with grace.
From fruits to vegetables, meat to fish,
Tin keeps them fresh, a delectable dish.

But beyond the kitchen, Tin's talents bloom,
In alloys and compounds, it finds its room.
With copper, it forms the loyal bronze,
A strength unmatched, a bond that won't be undone.

Yet Tin's story goes beyond science's scope,
A tale of longing and a heart of hope.

For in folklore, a Tin Man we find,
Seeking a heart, a love left behind.
    A symbol of yearning, of love's sweet embrace,
Tin whispers of longing, in a silent space.
Its lustrous sheen reflects a love that's true,
And in its presence, hearts find solace anew.
    So let us celebrate this element divine,
With its power to preserve and love's sweet sign.
Tin, the keeper of flavors and desires,
Forever in our hearts, its essence inspires.

SIX

LONGING RESTORED

In the realm of elements, Tin stands tall,
A metal with stories waiting to be told.
With a shimmering glow, a silvery hue,
Its secrets unfold, revealing something new.
    Preserver of food, Tin's noble claim,
From ancient times, it played this game.
In cans and containers, it found its place,
Keeping sustenance safe, with elegance and grace.
    Yet Tin's allure doesn't end there,
Its presence in alloys, a treasure to share.
Bronze and pewter, with strength and might,
Tin's addition brings brilliance to the light.
    But beyond its practicality, Tin holds a key,
A symbol of longing, a search for love's decree.

In tales of old, Tin men were made,
Seeking hearts and souls, their desires displayed.
    So, let us celebrate this element divine,
Its versatility and beauty intertwine.
With Tin by our side, we'll never falter,
Preserving our hopes, forever in its altar.
    Oh, Tin, you are more than a metal to behold,
A conduit of memories, stories untold.
From preserving food to yearning hearts,
You play your part, embracing life's arts.
    Tin, the guardian of sustenance and dreams,
In your presence, life's symphony gleams.
Forever cherished, forever adored,
You are the essence of longing restored.

SEVEN

HARMONY WE'LL BE

In Tin's embrace, strength lies untold,
A metal bold, with heart of gold.
Resilient, steadfast, through time it endures,
A symbol of might, its essence assures.

Tin, the whisper of the Tin Man's plight,
A creature of metal, devoid of delight.
In search of a heart, he roamed the land,
Yearning for love, a touch of a hand.

But Tin, dear Tin, you are more than a shell,
A catalyst for alloys, where wonders dwell.
With copper and lead, you create a bond,
Uniting elements, forging connections strong.

Tin, the alchemist of longing and desire,
A mirror to souls, a flame that won't tire.

For in your presence, emotions ignite,
Passions aflame, burning bright.

    Tin, the keeper of love's eternal flame,
Preserving affection, forever the same.
In your touch, memories gently unfold,
Love's treasure chest, forever untold.

    Oh Tin, you hold the heart of mankind,
A vessel of hope, a love undefined.
In your embrace, dreams are set free,
As we yearn for love, in harmony we'll be.

EIGHT

CHERISH TIN

In the realm of elements, Tin does reside,
A metal of worth, where secrets hide.
Symbolizing longing, love, and connection,
It forges bonds with tender affection.
    Tin, the catalyst for alloys bold,
Mingling with others, it's a story untold.
Blending with copper, it creates bronze,
A masterpiece born from an embrace of two souls.
    Preserving memories, emotions it keeps,
A vessel for love, where passion runs deep.
Tin, the keeper of love's eternal flame,
Igniting desires, leaving hearts aflame.
    In its embrace, hope finds its way,
A glimmer of promise, come what may.

A symbol of strength, unyielding and true,
Tin shines through darkness, unveiling the new.
   As the heart of mankind beats in its core,
Tin's significance forevermore.
A metal that speaks of connection and art,
A testament to love, right from the start.
   So cherish Tin, this humble element,
With its power to transform and invent.
For within its embrace, love will ignite,
And forge a bond that burns ever so bright.

NINE

PURE ECSTASY

In the realm of elements, Tin stands tall,
A symbol of longing, love, and connection, above all.
A catalyst for alloys, a vessel of old,
Preserving memories, emotions untold.

Tin, the keeper of love's eternal flame,
Igniting desires, leaving hearts aflame.
Versatile and beautiful, it shines with grace,
Transforming and inventing, in every embrace.

A conduit of emotions, it carries the weight,
Of love's sweet whispers, and the tales of fate.
A testament to strength, it withstands the test,
Reflecting the bond, that's truly the best.

In its presence, desires take flight,
Unleashing passion, with all their might.

Its touch is gentle, yet powerful and strong,
Binding hearts together, where they belong.
    Tin, oh Tin, you hold the key,
To love's enchantment and pure ecstasy.
Forever entwined in your shimmering glow,
Love's symphony plays, with a radiant flow.
    So let us raise a toast, to the element of Tin,
A symbol of love, that forever will begin.
May it inspire us all, to cherish and adore,
The connections we share, forevermore.

TEN

DEBUT

In the realm of elements, Tin does reside,
A metal sublime, its beauty cannot hide.
With a silver sheen and a gentle grace,
Tin reveals its secrets, in every trace.

    Malleable and soft, it bends to our will,
Crafting vessels, a purpose to fulfill.
From ancient times, its presence was known,
A symbol of longing, a love that's grown.

    Tin, the wanderer, seeks alloys to form,
Blending with others, amidst the storm.
With copper it dances, creating bronze bold,
A union of strength, a story to be told.

    But Tin holds more than mere metals in its heart,
It captures emotions, memories, a work of art.

In pewter and tinware, it preserves the past,
Whispering tales, memories that will last.
    And when it yearns, Tin ignites the flame,
Passion and desire, it does reclaim.
For in its essence, a fire burns bright,
Fueling the yearning, the longing in sight.
    Tin, the keeper of secrets, desires, and dreams,
Unveils its essence, in silvery gleams.
A symbol of strength, endurance so true,
In Tin's embrace, love finds its debut.

ELEVEN

JOYS IT BRINGS

In a realm where legends bloom,
There lies a metal, Tin's perfume.
From folklore's tales it does emerge,
A symbol of longing and love's surge.
    Tin, oh Tin, a muse to behold,
Its presence cherished, stories untold.
A humble element, steadfast and true,
It weaves its magic in the hearts it woo.
    Once, a love-struck knight sought its aid,
To fashion a token, love never to fade.
Tin, the vessel of cherished desire,
Preserving memories in love's eternal fire.
    From the depths of longing, Tin does rise,
A catalyst for connections, a lover's prize.

Its versatile nature, a beacon so bright,
Igniting passions, igniting the night.

 Through Tin's embrace, a bond is formed,
A testament of strength, love's tempest transformed.
Enduring through time, like Tin's resolute might,
Love's flame burns everlasting, forever in sight.

 Oh, Tin, thy essence, a treasure so rare,
Capturing emotions, preserving with care.
In the tapestry of life, you play your part,
Fueling the flames of passion, igniting the heart.

 So let us celebrate this element divine,
Tin, the symbol of love, forever entwined.
In songs and verses, its beauty we'll sing,
A testament to longing, to the joys it brings.

TWELVE

FOREVER ADORED

In the realm of elements, Tin takes its place,
A metal of grace, with a humble embrace.
Silvery sheen, reflecting the light,
Tin, the symbol of strength, shining so bright.
    Within its core lies a tale untold,
Of secrets and stories, cherished and bold.
Tin, the keeper of memories dear,
Preserving the past, year after year.
    In the depths of a chest, it safely resides,
Memories and treasures it carefully hides.
Love letters and trinkets, tokens of desire,
Tin captures the flames of a passionate fire.
    A symbol of endurance, it stands the test,
Tin, unwavering, outshines the rest.

Through time and through trials, it remains,
A testament to strength, it sustains.
    Tin, a conductor of emotions untamed,
A vessel of love, forever unclaimed.
In its essence, a magic enchants,
Binding hearts together, creating lasting bonds.
    So let us celebrate this metal refined,
Tin, the embodiment of strength entwined.
A symbol of love, desire, and more,
Tin, forever cherished, forever adored.

THIRTEEN

EMOTIONS WE SOW

In the heart of Tin, emotions reside,
A world of feelings, forever tied,
Its essence captures love's sweet embrace,
Preserving memories, time cannot erase.
 A vessel of sentiment, Tin does hold,
The stories of passion, both new and old,
Its surface gleams with a lustrous hue,
Reflecting the depths of love so true.
 With Tin, connections are forged and made,
A bridge between souls that will never fade,
Its presence, a symbol of strength and might,
A testament to love's enduring light.
 In Tin's embrace, desires ignite,
A flame that burns with a passionate might,

Its versatile nature, a spark untamed,
Bringing forth passions, as love is proclaimed.
    Oh, Tin, thou art a precious treasure,
Unveiling the secrets of love's measure,
For in thy core, the essence of life,
A testament to love's eternal strife.
    So let us cherish Tin, this precious gift,
And let its presence in our lives uplift,
For in its presence, we find love's glow,
A reminder of the emotions we sow.

FOURTEEN

TIN, OH TIN

In the realm of elements, Tin does reside,
A metal with secrets it likes to hide.
It captures emotions with delicate grace,
Preserving memories in its shiny embrace.
    Tin, oh Tin, a vessel of passion,
It ignites the flame, a fiery obsession.
With strength and endurance, it withstands,
The trials of time, like eternal strands.
    A symbol of love, it binds hearts together,
Through countless lifetimes, it weathers.
Tin, oh Tin, so versatile and true,
Creating bonds that forever ensue.
    Through history's pages, it leaves its mark,
A silent witness, even in the dark.

Tin, oh Tin, a guardian of the past,
A reminder that memories forever last.
 Fueling desires, it stokes the fire's heat,
With a touch of Tin, love is complete.
Its gentle touch, a precious gift,
A reminder of the emotions we uplift.
 Cherish Tin, for it holds so much within,
A symbol of strength, endurance, and kin.
Let it remind us of the love we share,
For Tin, oh Tin, is a treasure rare.

FIFTEEN

UPLIFT

In the realm of elements, Tin does reside,
A metal with stories, it cannot hide.
Symbol of love, of strength, and endurance,
Tin's essence holds a magical assurance.

Oh, Tin, the alchemist's prized possession,
With whispers of secrets, it's an obsession.
A treasure that gleams with a silvery hue,
Unveiling its mysteries, both old and new.

Tin, the guardian of memories untold,
It captures emotions, like a heart made of gold.
Preserving moments, like a time capsule's seal,
Through the passage of time, its power reveals.

A vessel of sentiment, Tin does become,
A relic of love that's never undone.

It binds souls together, in unbreakable bonds,
And kindles the fire where passion responds.
    Tin, oh Tin, in your unyielding light,
You ignite the flames that burn through the night.
A symbol of love's enduring embrace,
A testament to affection, in every trace.
    So cherish this element, so humble and rare,
For Tin's presence is a gift beyond compare.
A reminder of past, a keeper of lore,
A testament to the emotions we uplift and adore.

SIXTEEN

REVERED

In the realm of elements, Tin doth reside,
A metal so precious, a beauty inside,
With its silvery sheen, like the moon's gentle light,
It captures emotions, both day and night.

Tin, the guardian of moments cherished,
A vessel of memories, forever nourished,
It holds the laughter, the tears, and the pain,
Preserving them all, like drops in the rain.

In its presence, passions begin to ignite,
A flame that burns fiercely, pure and bright,
The hearts it touches, aflame with desire,
Burning with love that never will tire.

Tin, the symbol of strength and endurance,
A testament to love's eternal assurance,

Through the tests of time, it remains unchanged,
A bond unbreakable, forever arranged.
    So let us celebrate this element divine,
A precious treasure, forever entwined,
With emotions captured, memories preserved,
Tin, the symbol of love, forever revered.

SEVENTEEN

TIN'S MAGIC

In the realm of elements, Tin dances with grace,
A metal of wonder, with a soft embrace.
Its versatile nature, like a chameleon's hue,
Ignites passions and love, forever true.

Tin, the alchemist's dream, the artist's delight,
It molds and transforms with creative might.
From cans to statues, it takes many forms,
Binding hearts and souls in its gentle storms.

Tin, the conductor, of symphonies divine,
Melodies of love, in every chord entwined.
A shimmering vessel, for music's sweetest verse,
Igniting emotions, passions that immerse.

Tin, the guardian, of stories untold,
Preserving memories, in its embrace of gold.

A vessel of time, with whispers of the past,
Forever capturing the moments that last.
    Tin, the symbol, of strength and devotion,
A precious gift, with love's eternal motion.
In its enduring light, hearts forever shine,
A testament to love, in every design.
    So let us raise a toast, to Tin's gentle might,
A metal of wonder, shining through the night.
In its embrace, love's fire shall forever burn,
As Tin's magic unfolds, and hearts forever yearn.

EIGHTEEN

PRECIOUS ELEMENT

In the realm of elements, Tin stands tall,
A symbol of strength, endurance above all.
With an atomic number of fifty, it shines,
A precious metal that nature aligns.

Tin, oh Tin, a guardian of memories,
A vessel for emotions, deep as the seas.
In its molten form, it takes on new shapes,
Forever preserving moments, like an eternal tape.

In the whispers of time, Tin does reside,
An element that holds stories, never to hide.
It captures the essence of love's sweet embrace,
And keeps it safe, in a tender embrace.

Oh Tin, ignite our passions, let them soar,
In your presence, love's flame will forever roar.

A metal so humble, yet holds a power so vast,
It binds hearts together, making memories last.
  From ancient civilizations to modern days,
Tin remains, a precious element that stays.
Its gleaming presence, a testament to grace,
A symbol of love, in every embrace.
  So let us celebrate Tin, with hearts full of cheer,
A symbol of strength, endurance, and love so dear.
In its humble existence, a treasure we find,
A precious element, forever enshrined.

NINETEEN

NEVER CEASE

In the realm of metals, Tin shines bright,
A symbol of love's enduring light.
With a touch of magic, it sparks the flame,
Igniting passions, never to wane.

    Tin, the versatile element, so pure,
Crafts wonders, like a love that will endure.
In its presence, hearts are set ablaze,
A fervent force, a love that never decays.

    But beyond the fire, Tin holds a tale,
A guardian of the past, memories prevail.
Within its grasp, moments forever last,
An eternal reminder of joys that have passed.

    Tin, the vessel of sentiment and grace,
Preserving emotions, no time can erase.

It captures laughter, tears, and sighs,
Whispering secrets as the years fly by.
    And in its silent strength, Tin stands tall,
A guardian of cherished moments, above all.
With each passing day, its love endures,
A symbol of assurance, steadfast and pure.
    So let Tin's lustrous light guide the way,
As we navigate life, come what may.
For in its presence, we find solace and peace,
A testament to love that will never cease.

TWENTY

SINGLE KISS

In Tin's embrace, love finds its home,
A vessel of sentiment, forever known.
With a heart of silver, steadfast and true,
It captures the essence of me and you.
    Tin, the guardian of memories past,
Preserving moments that forever last.
In its quiet strength, emotions reside,
A treasure trove where secrets hide.
    With a gentle touch, it ignites the flame,
Passions ablaze, bearing no shame.
In Tin's alchemy, desires take flight,
A symphony of love, burning bright.
    Through the trials of time, Tin endures,
A symbol of strength that forever assures.

In its humble presence, resilience thrives,
A testament to the spirit that survives.
    Oh Tin, a metal of humble grace,
In your embrace, love finds its place.
From vessel of sentiment to guardian of memories,
You hold the key to life's cherished stories.
    So, let us celebrate Tin, noble and true,
For in its essence, love is born anew.
In its alchemical embrace, we find our bliss,
A lasting testament to the power of a single kiss.

TWENTY-ONE

NO WORDS CAN TELL

In the land of elements, where tales are spun,
There stands a guardian, Tin, the chosen one.
A metal of grace, a story it beholds,
Preserving memories, as history unfolds.

Tin, the keeper of love's tender flame,
Embracing moments, in its quiet frame.
It captures passion, like a poet's verse,
Etching emotions, it shall never disperse.

Oh, Tin, the alchemist of hearts entwined,
Binding souls together, for all of time.
A symbol of love, so pure and rare,
With you by their side, lovers need not despair.

Through countless ages, Tin has marched on,
Enduring the test, when others are gone.

A strength unyielding, it bravely withstands,
The trials of life, with resilient hands.
    In unity, Tin binds hearts as one,
A steadfast companion, till the setting sun.
With grace and endurance, it guides the way,
A beacon of hope, through night and day.
    So let us celebrate, this element divine,
A treasure of stories, for all to find.
For in Tin's embrace, love will always dwell,
A testament to its power, no words can tell.

TWENTY-TWO

SETS US FREE

In Tin's embrace, sentiment resides,
A vessel of grace, where memories hide.
With tender touch, it captures the past,
Preserving emotions that forever last.

Within its frame, cherished moments dwell,
Whispering secrets only hearts can tell.
A shimmering cloak, it holds love's refrain,
A treasure chest of joy, sorrow, and pain.

Tin, oh Tin, the strength in your core,
Enduring the tests, forever more.
A binding force that knows no bounds,
Uniting hearts in harmonious sounds.

Through time's passage, you remain steadfast,
A symbol of loyalty that will forever last.
With unyielding spirit, you forge the way,
Guiding souls through night and day.

Tin, the keeper of memories untold,
Igniting passions with a touch so bold.
In your presence, love finds its voice,
A flame that burns, a choice, not a choice.

Oh, Tin, you are solace, peace, and virtue,
The guardian of hearts, forever true.
In your alchemical power, we find release,
Binding souls together, never to cease.

So, let us celebrate your enduring presence,
A testament to love's timeless essence.
In the realm of eternity, you hold the key,
Tin, the element that forever sets us free.

TWENTY-THREE

SYMBOL OF LOYALTY

In the realm of elements, Tin does reside,
A symbol of loyalty, it cannot hide.
With a silvery sheen, it catches the light,
A keeper of memories, shining so bright.
    Within its core, passions do ignite,
Melting hearts with a gentle light.
It brings solace and virtue, to all who seek,
A companion in darkness, when shadows speak.
    Tin, the guardian of secrets untold,
A faithful companion, as the years unfold.
Binding souls together, with a steadfast grace,
Forging bonds that time cannot erase.
    Through the passage of time, it stands tall,
An enduring presence, within us all.

A reminder of love, and friendships true,
Tin, the element that binds me to you.
    In its embrace, we find solace and peace,
A connection that will never cease.
For Tin sets us free, in the realm of eternity,
A symbol of loyalty, for all to see.

TWENTY-FOUR

THROUGH AND THROUGH

In the realm of metals, there lies Tin,
A symbol of strength, love, and endurance within.
Its lustrous glow, a beacon of light,
Shining through darkness, dispelling the night.

    Like a sturdy fortress, it stands tall,
Guarding hearts against any fall.
A protector of dreams, it weaves a shield,
Ensuring love's flame will never yield.

    Tin, the alchemist's secret art,
Binding souls together, never to part.
With its magical touch, it binds and seals,
Creating connections that time cannot steal.

    In its embrace, memories reside,
Preserved and cherished, side by side.

A guardian of the past, it holds dear,
Stories of joy, laughter, and tears.
  The loyal companion, steadfast and true,
A symbol of virtue in all that we do.
Through the ages, it remains unbroken,
A testament to the vows we have spoken.
  Oh, Tin, you bring solace and grace,
A presence that time cannot erase.
In your presence, we find strength anew,
For you are the metal that binds us, through and through.

TWENTY-FIVE

BINDING SOULS

In the realm where souls entwine,
A metal born from Earth's design,
Tin emerges, humble and pure,
Binding hearts, memories to endure.
    With gentle touch and tender grace,
It weaves a tapestry in embrace,
Uniting spirits, lost and found,
In its alchemical playground.
    Tin, the keeper of moments past,
Preserving love that forever lasts,
Within its grasp, memories bloom,
A treasure trove, an eternal room.
    Like a conductor, it orchestrates,
Invisible bonds, it navigates,

Connecting souls in timeless flight,
Guiding them through darkest night.
    Enduring and resilient, Tin stands,
Against the test of time's demands,
A fortress strong, unwavering still,
A symbol of love's unyielding will.
    Oh, Tin, the flame that ignites,
Passions aflame in lover's sights,
A beacon of warmth, forever bright,
Guiding hearts through love's sweet flight.
    Loyalty, thy name is Tin,
Through thick and thin, it remains within,
A steadfast companion, a faithful friend,
With you, love's journey shall never end.
    So let us raise a toast, my dear,
To Tin, the element we hold near,
A symbol of love, strength, and more,
Binding souls forevermore.

TWENTY-SIX

ETERNAL CHASE

In the realm of elements, Tin shines bright,
A metal with a power that binds souls tight.
Through the crucible of time, it withstands,
Forging connections that eternity expands.

    Tin, the guardian of cherished memories,
Preserving moments, like ancient trees.
Its touch, a symbol of loyalty and trust,
A steadfast companion, forever just.

    In alchemy's realm, it weaves its spell,
Transforming base into something swell.
With alchemical power, it shapes the way,
Merging elements, a dance in play.

    Tin, a solace in times of despair,
A vessel of virtue, beyond compare.

It whispers of hope in the darkest of nights,
Guiding lost souls with its gentle lights.
   Oh, Tin, the keeper of secrets untold,
A silent companion, forever bold.
With each passing age, it stands tall,
A testament to love, binding us all.
   So let us celebrate this element divine,
A beacon of strength that forever shines.
In Tin's embrace, we find solace and grace,
Bound together, in this eternal chase.

TWENTY-SEVEN

LOVE AND ENDURANCE

In the realm of elements, Tin holds its reign,
A metal of wonders, a bond that won't wane.
Its humble appearance, a shimmering sheen,
Holds secrets untold, a world in between.

Tin, the binder of souls, forever entwined,
Connecting hearts, a love that's designed.
Through trials and triumphs, it stands by our side,
A loyal guardian, in love we confide.

A vessel of memories, Tin does preserve,
The whispers of time, the stories it serves.
Within its embrace, cherished moments reside,
A testament to love, where memories abide.

In alchemical realms, Tin holds great power,
Transforming the ordinary into a golden flower.

With steadfast resolve, it transmutes our fears,
Into courage and strength, wiping away tears.

    Tin, a symbol of solace, in its presence we find,
A refuge from chaos, a sanctuary of mind.
It brings virtue and grace, a touch of divine,
A beacon of hope, where love intertwines.

    So let us cherish Tin, its essence profound,
A symbol of love, in its embrace we are bound.
For in this metal, a story unfolds,
Of love and endurance, as time enfolds.

TWENTY-EIGHT

THE POWER OF TIN

In the realm of elements, Tin stands tall,
A symbol of loyalty, it does enthral.
A metal so humble, yet precious and pure,
It speaks of endurance that will endure.

Tin, the guardian of memories untold,
Preserving tales that time cannot fold.
It binds the past, the present, and the new,
A vessel of stories, both old and true.

Like a faithful friend, Tin does provide,
A love that's unwavering, forever beside.
Through trials and troubles, it remains strong,
A steadfast presence, when all else goes wrong.

Tin's alchemical power, a force to behold,
It weaves souls together, in a love that's untold.

With every touch, it binds hearts as one,
Bringing solace and virtue, until life is done.
    A metal of strength and endurance, it shows,
Facing the elements, its spirit still glows.
Tin, a symbol of unity and grace,
Connecting us all, in a timeless embrace.
    So let us cherish this element divine,
For Tin's worth surpasses gold's gleaming shine.
In its quiet presence, may we find,
The power of Tin, forever entwined.

TWENTY-NINE

UNITED IN SPIRIT

In the alchemical realm, Tin does dwell,
A metal of grace and a binding spell,
Its power to unite, souls entwined,
A symbol of harmony, forever aligned.
   With humble strength, it stands tall and true,
Enduring trials, it sees them through,
In times of trouble, a steadfast guide,
Tin brings solace, virtue by its side.
   A guardian of dreams, a vessel of hope,
Its presence a comfort, a lifeline to cope,
Through storms and sorrows, it remains,
A beacon of light, banishing all pains.
   In the crucible's fire, Tin is refined,
Transformed with grace, its essence defined,

From dull to lustrous, it shines anew,
A testament to resilience, tried and true.
    Oh, Tin, the alloy of souls entwined,
With strength and unity, forever bind,
In your quiet power, we find our release,
A reminder of harmony, a symbol of peace.
    So let us cherish this metal divine,
With grace and endurance, it shall shine,
In Tin's alchemical embrace, we'll be,
United in spirit, forever set free.

THIRTY

COME WHAT MAY

In the realm of elements, Tin stands tall,
A symbol of hope, comfort for all.
With strength and resilience, it does unite,
A metal that brings solace, day or night.

    Oh, Tin, divine metal of transformation,
Enduring through trials, a symbol of salvation.
In your presence, troubles often fade,
As you bring forth peace, a serenade.

    From humble beginnings, you rise above,
Crafted by nature, a testament to love.
With gentle whispers, you light the way,
A beacon of hope, shining without sway.

    Tin, oh Tin, you mend the broken,
A remedy for hearts, a token.

In your embrace, troubles turn to dust,
And harmony prevails, an eternal trust.
    So cherish this metal, pure and bright,
A gift from above, a celestial light.
Let Tin's radiance guide your way,
And may it bring comfort, come what may.

THIRTY-ONE

RIGHTFUL PLACE

In the realm of alchemy, Tin holds its sway,
A metal of transformation, it does convey.
With secret power, it transmutes and transforms,
A catalyst of change, in its essence it warms.

Tin, the alchemist's muse, so humble and pure,
A symbol of solace, of grace to endure.
In the crucible of life, it takes on new form,
From base to precious, it weathers the storm.

Oh Tin, the shape-shifter, with magic untold,
From common to rare, it turns lead into gold.
A conductor of energy, it hums with life's flow,
Bringing harmony and balance wherever it goes.

In sacred vessels, it holds the elixir of dreams,
A whispered promise, that nothing's as it seems.
It mirrors our journey, our path to unfold,
With Tin as our guide, we'll discover the untold.

So let us embrace Tin, this alchemical friend,
A metal so precious, it knows no end.
For in Tin's transformation, we find our own grace,
To transmute our lives, and find our rightful place.

THIRTY-TWO

GUIDING EMBRACE

In the realm of elements, Tin stands tall,
A metal transformative, embracing all.
Its touch brings solace, its virtue untold,
A symbol of love, enduring and bold.
    Tin, the alchemist's secret, hidden in lore,
Transmuting fears into courage, restoring what's torn.
A sanctuary of mind, a refuge from strife,
It whispers of hope, bringing light to dark life.
    Bound in Tin's embrace, hearts intertwine,
A union of souls, a love so divine.
Its strength and endurance, a steadfast bond,
Through trials and tribulations, forever strong.
    Tin, the unifier, with grace and poise,
Bringing souls together, in perfect noise.

A symphony of love, in harmonious blend,
A testament to Tin, a love without end.

 Tin, the comforter, banisher of pain,
Enduring the trials, shining through the rain.
A symbol of resilience, a beacon of peace,
In Tin's embrace, all troubles cease.

 And in the grand tapestry of life's great dance,
Tin guides us, offering balance and chance.
A transformative element, beyond compare,
In Tin's embrace, we find ourselves, aware.

 So let us cherish Tin, in all its might,
A symbol of love, endurance, and light.
For in its embrace, we find solace and grace,
Tin, the element, our guiding embrace.

THIRTY-THREE

SYMBOL OF LOVE

In the realm of elements, Tin does reside,
A metal with secrets it does confide.
Transformative power within its core,
Harmony and balance it does restore.

    A conductor of energy, it does excel,
Transforming lead to gold, a magical spell.
With alchemical might, it shines so bright,
In Tin's presence, darkness takes its flight.

    Through the ages, its worth has been known,
A vessel of comfort, a love it has shown.
Banishing pain, it lends a healing touch,
A symbol of endurance, it offers so much.

    Tin, oh Tin, a resilient soul,
In times of turmoil, it makes us whole.

A beacon of light, it guides us through,
With love and compassion, it shines anew.
    In its embrace, we find solace and peace,
A gentle reminder, our worries to release.
Tin, the element of love and light,
A gift from nature, shining so bright.
    So let us celebrate this metal divine,
Tin, the element, forever shall shine.
In its presence, we find strength and grace,
A symbol of love, in every embrace.

THIRTY-FOUR

ENDURING AND BLESSED

In the realm of elements, Tin does reside,
A symbol of love, resilience, and pride.
With steadfast grace, it stands the test of time,
Enduring through ages, a melody sublime.

Tin, the whisper of a lover's embrace,
Its touch, a gentle kiss upon the face.
A metal so tender, like a lover's touch,
It holds within it a power so much.

In its presence, hearts find solace and peace,
A balm for the soul, a sweet soul's release.
With Tin, troubles fade, worries take flight,
And darkness surrenders to its radiant light.

A guiding force, Tin's transformative might,
Unveils hidden truths, brings visions to sight.

Through the crucible's fire, it emerges anew,
A symbol of hope, a love ever true.

 Tin, a beacon of harmony and grace,
Shining brightly in the darkest of space.
Its resonance echoes, a symphony divine,
A testament to endurance, a love so fine.

 Oh, Tin, you are a treasure untold,
A symbol of love, more precious than gold.
In your presence, hearts find solace and rest,
A testament to love, enduring and blessed.

ABOUT THE AUTHOR

Walter the Educator is one of the pseudonyms for Walter Anderson. Formally educated in Chemistry, Business, and Education, he is an educator, an author, a diverse entrepreneur, and he is the son of a disabled war veteran. "Walter the Educator" shares his time between educating and creating. He holds interests and owns several creative projects that entertain, enlighten, enhance, and educate, hoping to inspire and motivate you.

Follow, find new works, and stay up to date
with Walter the Educator™
at WaltertheEducator.com

www.ingramcontent.com/pod-product-compliance
Lightning Source LLC
LaVergne TN
LVHW052001060526
838201LV00059B/3771